ANIMALS OF THE TUNDRA

ROAMING CARIBOU

THERESA EMMINIZER

I0811099

PowerKiDS press

New York

Published in 2022 by The Rosen Publishing Group, Inc.
29 East 21st Street, New York, NY 10010

Copyright © 2022 by The Rosen Publishing Group, Inc.

All rights reserved. No part of this book may be reproduced in any form without permission in writing from the publisher, except by a reviewer.

First Edition

Portions of this work were originally authored by Roman Patrick and published as *Caribou*. All new material in this edition authored by Theresa Emminizer.

Editor: Jill Keppeler
Book Design: Michael Flynn

Photo Credits: Cover, p. 1 Daniel A. Leifheit/Moment Open/Getty Images; (series background) Stefan Sorean/Shutterstock.com; p. 5 Mark Newman/The Image Bank/Getty Images; p. 7 Tom Walker/Stockbyte/Getty Images; p. 9 (inset) Mark Daffey/Lonely Planet Images/Getty Images; p. 9 (main) aaabbbccc/Shutterstock.com; p. 11 (inset) Rumo/iStock/Getty Images; p. 11 (main) Tom Walker/Photographer's Choice/Getty Images; p. 13 Westend61/Getty Images; p. 15 Geoffrey Reynaud/iStock/Getty Images; p. 17 Patrick J. Endres/Corbis Documentary/Getty Images; p. 19 Dennis Stogsdill/iStock/Getty Images; p. 21 Iakov Filimonov/Shutterstock.com.

Library of Congress Cataloging-in-Publication Data

Names: Emminizer, Theresa, author.
Title: Roaming caribou / Theresa Emminizer.
Description: New York : PowerKids Press, [2022] | Series: Animals of the tundra | Includes index.
Identifiers: LCCN 2020021709 | ISBN 9781725326477 (library binding) | ISBN 9781725326453 (paperback) | ISBN 9781725326460 (6 pack)
Subjects: LCSH: Caribou–Juvenile literature. | Tundra animals–Juvenile literature.
Classification: LCC QL737.U55 E55 2022 | DDC 599.65/8–dc23
LC record available at https://lccn.loc.gov/2020021709

Manufactured in the United States of America

Some of the images in this book illustrate individuals who are models. The depictions do not imply actual situations or events.

CPSIA Compliance Information: Batch #CSPK22. For Further Information contact Rosen Publishing, New York, New York at 1-800-237-9932.

Find us on

CONTENTS

By Any Other Name

What's the difference between a reindeer and a caribou? There isn't one! In North America, people call wild reindeer by the name "caribou." These big animals live in the **tundra** and forests far to the north.

What Are Antlers?

Both **male** and **female** caribou have antlers. These are bony, branched growths on an animal's head. Male caribou may fight each other with their antlers. Males can be up to 5 feet (1.5 m) tall at the shoulder!

Brown and White

Caribou have a coat of hairs that are hollow, or open in the middle. This traps air and keeps them warm. Some caribou have a brown coat all year. Some turn white in the winter. They match the snow around them.

Hard Hooves

Caribou have special feet to help them walk on deep snow. Their big hooves are split to spread out on the snow. The hooves also help caribou swim and dig for food.

On the Move

Caribou may travel thousands of miles on yearly **migrations**. They move to southern forests to find food to eat in winter. A caribou can eat 12 pounds (5.4 kg) of food a day!

Huge Herds

Caribou move in huge groups called herds. A herd may have 500,000 caribou in it! In the spring, the caribou head back north. They use the same paths year after year.

Back to the Tundra

In the spring and summer, the tundra has more food for caribou to eat. They eat grass, mushrooms, and many other plants. Female caribou have their babies, or calves, in the summer.

Run Away!

Caribou have many **predators**, including wolves. When a caribou is scared by a predator, it puts its tail in the air, holds its head high, and runs! Caribou can run 48 miles (77.3 km) per hour.

Caribou and People

Sometimes people hunt caribou. Some native peoples keep **domesticated** caribou for food or milk. They may use their antlers to make tools. Sometimes these reindeer are trained to carry or pull loads or even carry riders!

Fancy Features

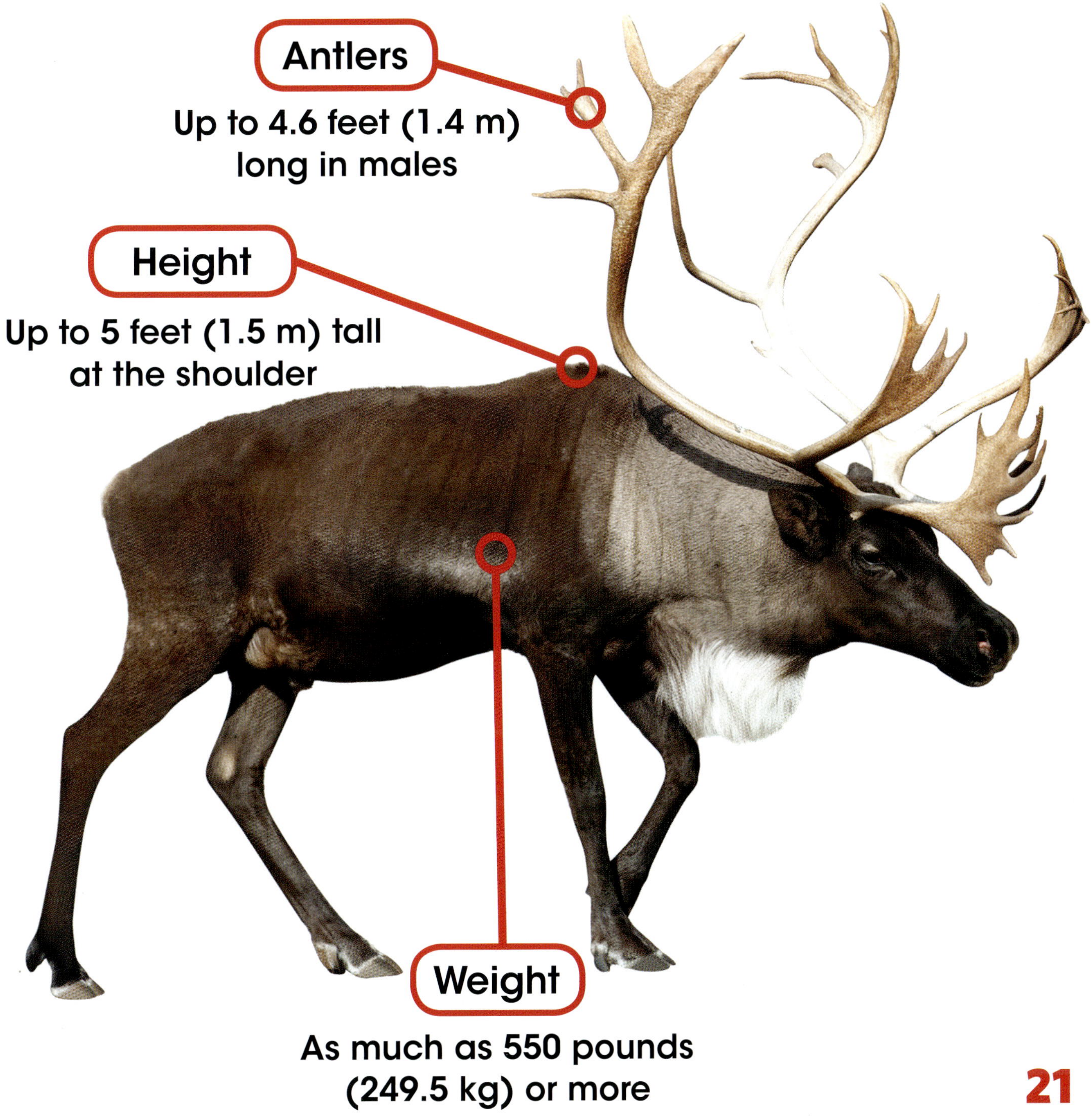

GLOSSARY

domesticated: Bred and raised for use by people.

female: A girl.

male: A boy.

migration: The act of an animal or people moving from one location to another.

predator: An animal that hunts other animals for food.

tundra: Cold northern lands that lack forests and have permanently frozen soil below the surface.

FOR MORE INFORMATION

WEBSITES

Caribou
kids.nationalgeographic.com/animals/mammals/caribou/
Learn more about caribou and take a peek at photographs of these fascinating creatures on this National Geographic Kids website.

Caribou: An Iconic Canadian Species
www.wwf.ca/conservation/arctic/wildlife/caribou/
Where do caribou live? How are they threatened? Learn the answers to these questions and more on this WWF website.

BOOKS

Borgert-Spaniol, Megan. *Caribou*. Minneapolis, MN: Bellwether Media, 2018.

Koestler-Grack, Rachel. *Caribou*. Mankato, MN: Amicus/Amicus Ink, 2019.

Publisher's note to parents and teachers: Our editors have reviewed the websites listed here to make sure they're suitable for students. However, websites may change frequently. Please note that students should always be supervised when they access the internet.

INDEX